# This Book Belongs To

___________________________

## Welcome to Your Math Adventure!

Hi there, Math Explorer!
This book is full of fun bugs who need your help with math!
Here's how to be the best bug helper:

### How to Complete Each Page:

Read or listen to the question carefully.

Take your time - there's no rush.

Use your finger to point and count if you need to.

Ask for help if you need it - that's what grown-ups are for.

Color in your star at the bottom when you finish - you earned it!

### Remember:

Every answer you try is helping you learn.

It's okay to make mistakes - that's how we get better!

Have fun with your bug friends!

**Look for Buzzy the Bee's challenges:** He will have some extra math adventures for you to try.

## Note for Parents and Caregivers

About This Book This workbook introduces essential pre-kindergarten math concepts through engaging bug-themed activities. Each section builds foundational skills that prepare children for future mathematical learning.

### Getting Started:

- Choose a quiet time when your child is alert and focused.
- Keep sessions short (10-15 minutes).
- Have crayons, pencils, scissors, and glue nearby.

### Your Role:

- Read instructions aloud if needed.
- Encourage counting out loud.
- Celebrate effort: "I love how carefully you counted!"
- If frustrated, try: "This is tricky, but your brain is working it out".

**About Complex Questions:** Some questions ask "Which one is NOT..." These build flexible thinking and prepare children for advanced reasoning. Help them look at all options first, then eliminate choices together.

**Extending Learning:** Connect to daily life by sorting laundry, counting during walks, and noticing patterns in tiles or nature.

**Remember:** Your patience and enthusiasm are most important. Math should feel like solving fun puzzles together!

# Table of Contents

# SORTING BY SHAPE & SIZE

Circle the biggest ladybug and put an X on the smallest.

**Buzzy's Challenge:**

How many ladybugs are in the picture?

"Way to go!"

## 2

Draw a line between each butterfly and flower that has the same shape.

**Buzzy's Challenge:**
How many petals does each flower have?

"Fantastic job!"

Circle the biggest beetle in each row.

Buzzy's Challenge:

How many legs does this beetle have?

"You rock!"

# 4

Look at each row of caterpillars.
Circle the two that are the same size.

**Buzzy's Challenge:**
How many antennae are on each caterpillar?

"Bug-tastic work"

Draw a line to connect the ants that are the same size.

Buzzy's Challenge:

Which ants are the smallest?

"You're a math star!"

6

Cut out or carefully tear the bees below the columns. Glue the big bees in the BIG BEES column and the little bees in the LITTLE BEES column.

| Big Bees | Little Bees |
| --- | --- |
| | |

Circle the firefly in each group that is not the same size.

**Buzzy's Challenge:**
How many fireflies are on the page?

"You did it!"

8

Color the leaves with the squares to help the spider get to its web.

**Buzzy's Challenge:**
Which shape appears least often on this page?

"High five!"

# Coloring Break

# Patterns

10

Do you see the pattern in each row?
Follow the pattern to color the last beetle.

**Buzzy's Challenge:**
How many red beetles are there in this activity?

"You're buzzing with knowledge!"

Circle the bug in the box that should come next in each row.

Buzzy's Challenge:

How many spiders appear on this page?

"Mission accomplished!"

Firefly Parade. Cut out or carefully tear the fireflies below the rows. Glue the correct color firefly to complete the pattern.

Draw a line to connect the bees that have the same pattern.

**Buzzy's Challenge:**
How many bees are on the page?

"Way to go!"

Find the butterfly in the garden that will complete the pattern and circle it.

**Buzzy's Challenge:**

What color would be next?

"Fantastic job!"

Draw the shape on the ladybugs' wings that completes the pattern in each row.

Buzzy's Challenge:
Which shape appears the most in this activity?

"You rock!"

16

Look at the pattern carefully to figure out which colors are missing, then color in the white circles with the correct colors.

**Buzzy's Challenge:**
Which color appears the most on the page?

"Bug-tastic work"

Help the grasshopper get to its lunch by following the pattern below.

**Buzzy's Challenge:**
How many grasshoppers are on this page?

"You're a math star!"

## 18

Circle the dragonfly that comes next.

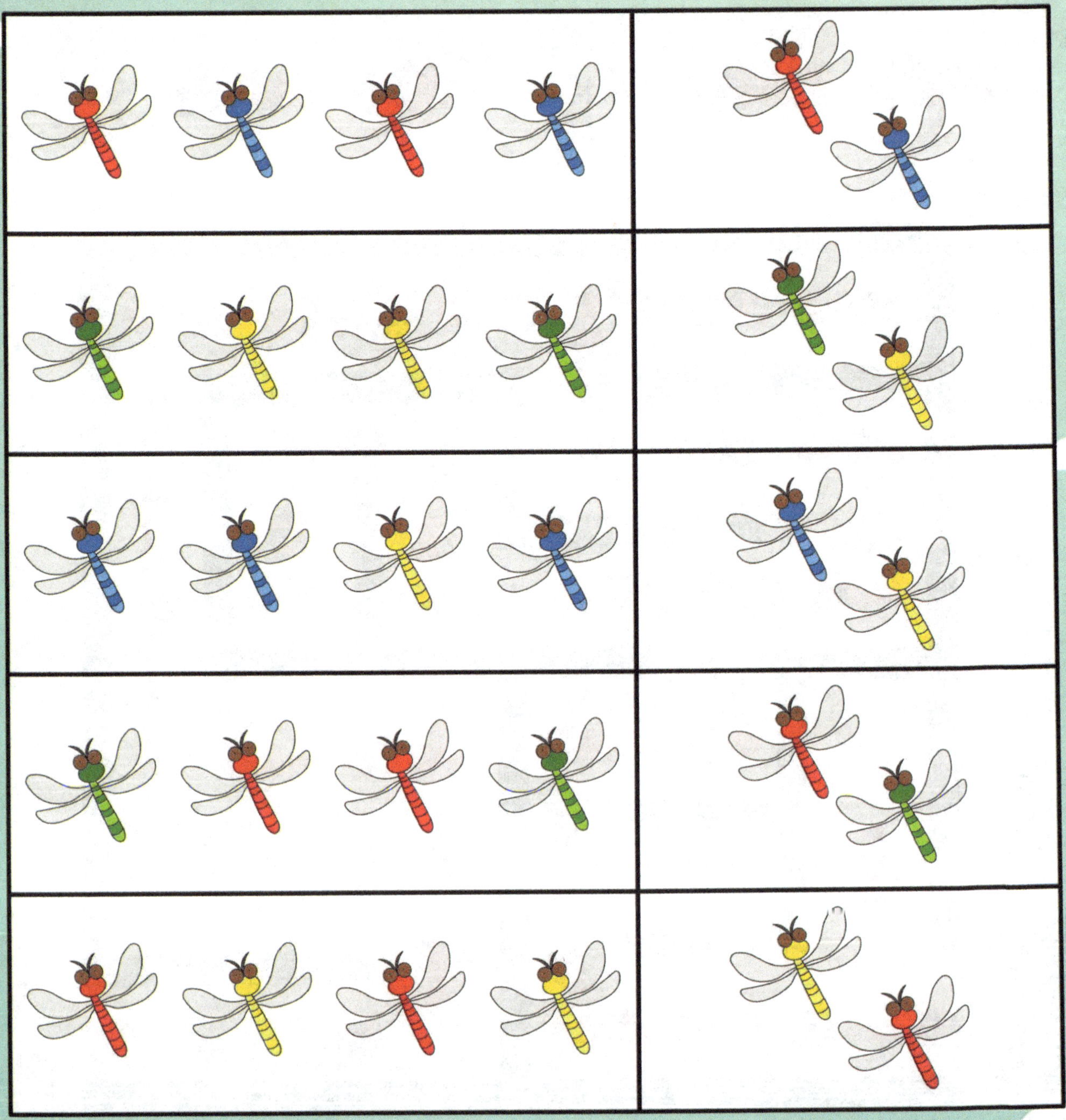

**Buzzy's Challenge:**

How many yellow dragonflies are on the page?

"You did it!"

Draw a line to connect the spider to the web with the same shape.

Buzzy's Challenge:
How many legs do the spiders have?

"High five!"

Make your own bug pattern.

**Buzzy's Challenge:**

Which bug did you use the most?

"Mission accomplished!"

# Coloring Break

# Position & Direction

Circle the butterflies flying above the flowers.

**Buzzy's Challenge:**
How many butterflies are flying below the flowers?

"High five!"

Put an X on the ladybugs under the leaves.
Circle the ladybugs on top of the leaves.

**Buzzy's Challenge:**
Which ladybug is the highest in the tree? Which ladybug is the lowest in the tree?

"Way to go!"

Circle the dragonflies next to the cattails.

**Buzzy's Challenge:**
How many dragonflies are above the cattails?

"Fantastic job!"

Help the bee get to the hive by taking it up and down in the maze.

**Buzzy's Challenge:**
How many bees are flying around the hive?

"You rock!"

Circle the ants crawling under the log.

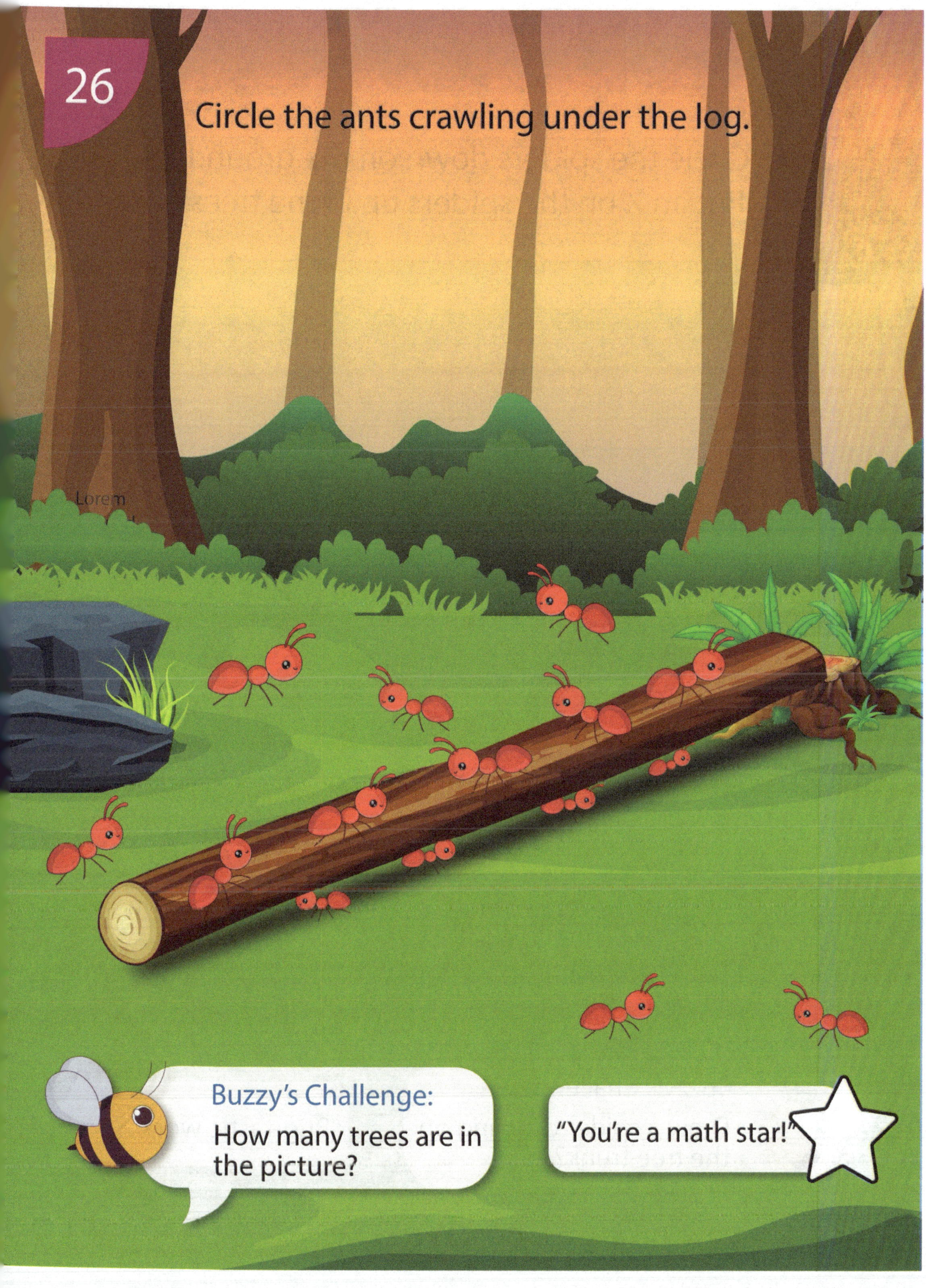


Buzzy's Challenge:

How many trees are in the picture?

"You're a math star!"

Circle the spiders down on the ground.
Put an X on the spiders up in the trees.

Buzzy's Challenge:
Draw a spider crawling on the tree trunk?

"Bug-tastic work"

Draw a line to connect the bugs in the same position.

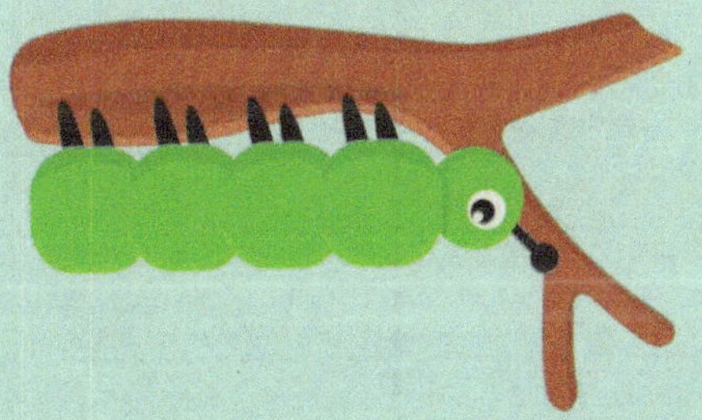

**Buzzy's Challenge:**

How many beetles are on this page?

"You're buzzing with knowledge!"

Trace the caterpillars' paths up and down.

**Buzzy's Challenge:**
Which caterpillar has a zigzag path??

"Mission accomplished!"

30

Cut out or carefully tear the butterflies and glue them so they are flying above the grass.

Circle the fireflies outside the jar.

**Buzzy's Challenge:**
Are there more fireflies inside the jar or outside?

"Way to go!"

# Coloring Break

# Counting

Count the spots on each ladybug.

**Buzzy's Challenge:**

Which ladybug has the fewest spots? Which ladybug has the most spots?

"Fantastic job!"

Count the grasshoppers and trace the numbers in each row.

1

2

3

4

5

**Buzzy's Challenge:**

Which number comes after 5?

"You rock!"

Put an X on the bees with 0 stripes.

**Buzzy's Challenge:**

How many bees have zero stripes?

"Bug-tastic work"

Draw a line to connect the groups that have the same number of ants.

**Buzzy's Challenge:**

Which group has neither the most or the fewest ants?

"You're a math star!"

37

Circle the flower that has 3 butterflies near it.

**Buzzy's Challenge:**
Put an X on the flower with zero butterflies?

"You did it!"

Circle the spiderweb in each row that has the same number of spiders.

**Buzzy's Challenge:**
How many spiderwebs have zero spiders?

"High five!"

## 39

Circle the jar with 5 fireflies.

**Buzzy's Challenge:**
Which jar has the least number of fireflies?

"Mission accomplished!"

Put an X on the lily pad with 1 dragonfly.

**Buzzy's Challenge:**
Which lily pad has the most dragonflies?

"You're buzzing with knowledge!"

Circle the leaves with 2 ladybugs.

**Buzzy's Challenge:**
What number comes before 5?

"Way to go!"

Find and circle 4 ladybugs.

**Buzzy's Challenge:**

How many bees are in the garden?

"Fantastic job!"

Color the correct number of circles for each caterpillar.

Color 4 circles.

Color 3 circles.

Color 2 circles.

Color 5 circles.

**Buzzy's Challenge:**

Which caterpillar has the fewest circles colored?

"You rock!"

Look at the dots in each section of the butterfly. Count the dots, then use the color key below to color that section.

1 dot = Black

2 dot = Red

3 dot = Green

4 dot = Blue

5 dot = Yellow

**Buzzy's Challenge:**

How many different crayons did you use to color the butterfly?

"You rock!"

## 45

Draw a line to connect the groups that have the same number of grasshoppers.

**Buzzy's Challenge:**
Which group has a number closest to 5?

"You're a math star!"

Count the bugs on each leaf.

**Buzzy's Challenge:**
Which leaf has the most bugs?

"You did it!"

# Coloring Break

# Comparing Size & Quantity

Circle the biggest bug in each row.

**Buzzy's Challenge:**
How many bugs in each row are the same size?

"Bug-tastic work"

Circle the smallest bug.

Buzzy's Challenge:
How many butterflies are the same size?

"You're a math star!"

Circle the jar that has more fireflies.

**Buzzy's Challenge:**
How many would each jar have if you added one more firefly?

"Mission accomplished!"

Which leaf has the fewest ladybugs.

**Buzzy's Challenge:**
How many would each leaf have,
if you added 1 more ladybug?

"You rock!"

Draw a line to connect the caterpillars that are the same size.

**Buzzy's Challenge:**

How many caterpillars are on this page?

"High five!"

53 Look at the picture. Circle the smallest butterfly. Put an X on the biggest bee. Underline the ant hill with the most ants.

Circle the dragonfly in each row that has the longest body.

Buzzy's Challenge:

Count the dragonflies in each row that have the shortest bodies?

"You rock!"

55

## Circle the beetle with the most spots.

**Buzzy's Challenge:**

How many beetles have exactly 3 spots?

"High five!"

Put an X on the spider web that
has the fewest spiders.

**Buzzy's Challenge:**
How many spiders would be on each web if you added one more?

"Fantastic job!"

Circle the logs that have the same number of ants.

**Buzzy's Challenge:**

How many ants are on each log?

"You're a math star!"

Put an X on the caterpillar that is the shortest in each row.

**Buzzy's Challenge:**
Which caterpillars are the same size in each row?

"You rock!"

Circle the hive with more bees.
Put an X on the hive with fewer bees.

**Buzzy's Challenge:**

Which hive would need the fewest bugs added to reach 10?

"Bug-tastic work"

# Coloring Break

# Certificate of Completion

Awarded to:

______________________________

Date:

______________

Great job!
Keep exploring
and learning!

Made in the USA
Monee, IL
02 November 2026

67379301R00044